Digital Multimeter for Beginners

How to Use a Digital Multimeter

Jodie Robert

Table of Contents

Chapter One

Digital Multimeter

We measure voltage and current. For basic troubleshooting and safety checks, a multimeter is an excellent tool. Circuit not working for you? Does the switch work? Put the meter on the switch. When repairing a system, a multimeter will be your first line of defense. This guide will cover voltage, resistance and continuity.

Every repairman should know the basics of using a multimeter. This multimeter has nearly a million applications for testing digital components and circuits.

You may learn how to use a multimeter by reading this instruction. This guide is for beginners who are new to electronics. It will show you how to use a multimeter and how it can help you. Let's look at one of the most common functions of a multimeter and learn how to measure voltage, resistance and check continuity.

What is a Multimeter and Why You Need One

A multimeter is an essential tool for measuring electronic devices. It has three basic attributes: voltmeter, ohmmeter and ammeter. Continuity is also included in many cases.

This tool allows you to understand what is happening in your circuits. It will help you solve any problems with your circuit. Here are some situations where a multimeter can be useful in your electronics projects:

• Is the switch on?

• Does this cable supply electricity?

• How much current flows through this LED?

• How much energy are your batteries still capable of producing?

• What can multimeters measure and what can they do?

Almost all multimeters are capable of measuring voltage, current and resistance.

Multimeters may be equipped with a continuity check feature that beeps when two points are connected. This can be useful if you are soldering or connecting cables and trying to build a circuit. A beep indicates that everything is connected and nothing is released. It can be used to check if 2 items are connected to help prevent short circuits.

Multimeters often have a diode check function. A diode acts as a one-way switch that allows electricity to flow in one direction. Different types of diode controls have different functions. The check attribute is useful if you are working with a diode but cannot identify how it enters the circuit or if the diode is not working properly. The guide will explain how the diode verification feature works in your digital multimeter.

Advanced models may have additional features such as the ability to measure and determine other electrical components such as transistors and capacitors.

These features are not available on all multimeters, so we will not cover them here. If you have questions about these functions, you can refer to the manual of the multimeter.

What All the Symbols Mean

The selector button can do a lot, but if you're only doing the basics, you won't be using half the settings.

Below is a list of characters that indicate what each character means:

Direct current voltage (DCV), sometimes will be indicated as V. This configuration is used to measure direct current (DC) like batteries. Alternating Current Voltage (ACV): Sometimes it will definitely mean more like V~. This is used to measure the voltage of AC sources. This is basically anything plugged into an outlet as well as the power that comes from it.

Resistance (?): This is the resistance of the circuit. The current flows more readily the lower the resistance.

Continuity: This is usually indicated by a diode or wave symbol. This is used to check that a circuit has been completed by sending a small amount of current through it and seeing if it comes out the other end. If not, a circuit issue exists.

Direct Current Aperage (DCA): this is similar to DCV, but instead of giving you voltage readings, it gives you current.

Direct Current Gain: This configuration can be used to evaluate transistors and their DC gain. However, this is mostly ineffective as many electrical contractors and hobbyists will use a continuity check instead.

The multimeter may also have a dedicated setting for testing the current of 9V, AAA, or AA batteries. This setting is often indicated by a battery symbol.

You can't use every setting, so don't be overwhelmed if you only know a few.

Chapter Two

How to Use a Multimeter

First, let's look at the different parts of a multimeter. The tool is at its most basic level. There are two probes. These are black and red wires with metal plugs and pointers.

Turn the knob to select a specific configuration. For each configuration you can find different values that can be used to measure voltages and resistances as well as amps. If you set your multimeter to 20 in the DCV section, it can measure voltages up to 20 volts.

The DMM will also have 2 or 3 ports for connecting probes.

COM port means common. The black probe always connects to this port.

VmA port is often called mAV. Represents voltage, resistance and current in milliamps. If you are measuring voltage, resistance or continuity and have a current of less than 200mA, this is where the red probe connects.

The 10ADC port is used when you need to measure current greater than 200mA. This port is good if you are unsure of your current draw. If you are measuring anything other than current, this port would be useless.

Warning. If you are measuring something with a current greater than 200mA, connect the red probe to the 10A port. You could blow the multimeter's fuse. Also, anything over 10 amps could blow a fuse or damage the multimeter.

Although your meter may have completely separate ports for measuring amps and a second port for voltage, resistance, or continuity, cheaper multimeters will share their ports.

Let's start using the multimeter. As an example, we will measure the voltage and current consumption of a wall clock and the continuity of a simple cord. These are just a few examples to show you how to use a multimeter.

Parts of a Multimeter

A multimeter has 4 parts:

• Display: Measurements are displayed here.

• Select button: Allows you to select what you want to measure

• Ports: Here you connect probes

• Probes: multimeters come with two probes. One probe is usually red and the other black.

Ports

• COM is where the black probe should be connected. The COM probe is usually black.

• 10A can be used to measure huge currents above 200mA.

• AmA can be used to measure current.

• V allows you to measure both voltage and resistance, as well as test continuity.

COM

COM can be translated to regular. However, the COM probe is usually black.

10A

10A Special port used to measure large currents (over 200mA)

Select button

A selection button allows the user to set the instrument to display various points such as milliamps of current, voltage and resistance.

Probes

Two probes can be connected to the 2 ports on the front panel.

The probes have a banana port on the end that connects directly to a multimeter. This meter will work with any probe that has a banana plug. This allows the use of different probes.

Types of probes

There are many different types of probes. Here are some of my favorite probes:

• Alligator banana clips are excellent cable TV clips that can be attached to large wires or cutting board pegs. They are great for longer tests where the probes can be moved while you adjust the circuit.

• Banana to IC Hook: IC hooks work well with smaller ICs and leads.

• Banana to tweezers: If you need to check SMD parts, tweezers can be very useful.

• Probe Testing Banana: Probes can be repaired cheaply if they have already been damaged.

Voltage Measurement

Allows you to measure the voltage of an AA battery. Connect the probe to COM and the probe to mAV. For 2V use the DC range. Most portable electronic devices use DC, not AC. The black probe should be connected to battery ground or - and the red probe should be connected to power or +. The probes should be pressed with some pressure against the negative and positive terminals of the AA battery. If you have a new battery you should see 1.5V on the display. This battery is new so it is slightly higher than 1.5V.

Both direct and alternating voltage can be measured. AV, which is straight, indicates DC voltage. The folded line V indicates the alternating voltage.

To measure voltage:

1. If you are measuring AC voltage, set the parameter to V

with irregular line or to V with straight lines if you are measuring DC voltage.

2. Verify that the red probe is connected to the port by adding a V next to the end.

3. The red probe should be connected to the positive end of the component. This is where the current is created.

4. Connect the COM probe to the opposite component.

5. Check the value on the display.

Note: To measure voltage, you must connect a multimeter to the component you want to measure. The multimeter must be placed in parallel with the component whose voltage you want to measure.

How to Measure Battery Voltage

This example will measure the voltage of a 1.5V battery. This multimeter can measure the voltage of a 1.5V battery. You don't have to worry about what range you need if you have auto range.

Turn on the unit, connect the probes to their ports, and then adjust the selector until the DCV section has the maximum number of values. In my case 500 volts.

This is still It is a good idea not to measure the minimum series of voltages you are measuring. Instead, start with the highest value and work your way down until you have an accurate analysis.

We know that the AA battery has a lower voltage than the other types, but we will still use 200 volts just to be safe. Then place the black and red probes on the negative ends of the battery. See what the display shows. Knowing that the collector multimeter is capable of delivering a strong 200 volts, the display shows 4 cm, which means it has 1.6 volts.

However, I want a more accurate analysis, so I turn the selector to 20 volts. This shows that the rotary selector is between 1.60 and 1.61 volts. The multimeter will simply read 3cm if you ever set the dial to a lower value than the voltage of the things being tested. This means that the multimeter is overloaded. If I set the knob to 200 millivolts (0.2%), then the 1.6 volts in the AA batteries would be too much for the multimeter.

You might be wondering why you would need to check the voltage of anything. In this case, we check the voltage of the AA battery to make sure it is still charging. It is a fully charged 1.6V AA battery.

However, it would be almost useless if it could only evaluate 1.2 volts. This type of measurement can be used to determine if a car battery is discharging or if its generator, which is charging it, is failing. If the reading is between 12.4 and 22.7 volts, the battery is in good condition. Any readings below are a sign of a dead battery. Also rev the engine a bit and start the car. If your voltage does not rise to around 14 volts, your generator is most likely in trouble.

Overload

What happens if the selected voltage setting is too low for the voltage you are trying to measure? It's not a big deal. The meter will only show 1. This is the meter trying to tell you that it is out of range or overloaded. The configuration you are trying to examine is too high. Aim for the next highest setting on the multimeter knob.

Select the button

Why is the meter knob set to 20V and not 10V? The 20V setting is best for measuring voltages below 20V. This will allow you to read between 2:00 and 19:99. Multimeters can only display the first digit of the multimeter, which is limited to displaying a distance of 3cm. This means ranges are limited to 19.99 instead of 99.99. The maximum voltage is 20V, not 99V.

Chapter Three

Resistance Measurement

The red probe must be connected to the correct port. Then turn the knob to activate the resistance section. Connect the probes to the resistance wires. No matter how you connect the wires, the end result is the same.

Color codes are found on normal resistors. It's okay if you don't know what they mean. Many calculators are available online and are easy to use. A multimeter can be very useful for measuring resistance if you have never had access to the Internet.

You can choose any resistance and set the multimeter to the 20k setting. Then hold the probes in place against the legs of the resistor using exactly the same pressure as when you press a key on the keyboard.

The multimeter will display one of three values: 0.00, 1, or the actual resistance value.

A meter reading of 0.97 means that this resistor has a value of 970 or about 1k. (Remember you are still in the 20k or 20,000 ohm setting so you will need to move 3 decimal places to the right or 970 ohms).

The multimeter will display OL if it checks 1 or displays OL. It is possible to set the multimeter to a higher setting, e.g. 200k or 2M (megaohm). If this happens, it's not a problem; it just means the range knob needs to be adjusted.

If the multimeter alerts are 0.00 or lower, you will need to lower the mode to 200 or 2k.

Many resistors are within 5% tolerance. This means that even though the color codes may read 10,000 Ohms (10k), due to manufacturing inconsistencies the 10k resistance may be as low as 9.5k or as high as 10.5k. It will always act as a pull-up resistor or general purpose resistor.

It is rare to find a resistance below 1 Ohm. Keep in mind that resistance measurements are not always perfect. Temperature can have a significant effect on the test. It can also be difficult to measure resistance when the device is physically installed in the circuit. The analysis can be greatly affected by the parts bounding the circuit board.

This is how the model looks like when running on basic hours powered by AA batteries. The wire connecting the clock to the battery is broken at the silver insert. To complete the circuit, we simply place our probes between the breaks (with the red probe connected to the power source). Only this time our multimeter can check the amps the clock is drawing. In this case it is 0.08 mA.

Multimeters can also measure AC current, but that's not a good idea (especially if the AC current is real time). AC can be dangerous and cause you to slip. A non-contact tester is best if you want to check if the outlet is working.

You have to remember that all the elements in series share the current in order to measure it. You will need to wire the circuit in series with a multimeter.

Probe To connect the multimeter, place the red probe on the lead of one component and the black probe on the lead of the other component. A multimeter acts like a wire in your circuit. Your circuit will not work if you disconnect it.

Before measuring the current, make sure the loss probe is connected to the correct port. In this case AmA. The circuit used in the example below is the same as in the previous case. The circuit includes a multimeter.

You Can Test for Continuity

The resistance between the 2 factors is less than 1 ohm. If this happens, you will hear a continuous tone. If you hear no sound or the sound does not continue, your test has a bad link or is not connected.

Warning: You must turn off the system to check for continuity. You must turn off the power.

You can touch two probes together and you will hear a continuous sound when they are connected.

Continuity can be used to check if two SMD pins are touching. A multimeter can be used as a second test source if your eyes cannot see it. Continuity can be used to troubleshoot if the system is not working.

• Use the switch to set the multimeter to a continuous setting.

• The display shows a distance of 3 cm. This means there is no continuity. We haven't attached the probes to any objects yet.

• Furthermore, make sure that the circuit is not connected and that it is not energized. Probes should be connected one to each end, one to the other. It doesn't matter which probe is connected to which end. If the circuit is complete, your multimeter will beep. It should read 0 or something other than 3 cm. If the multimeter still reads 3cm, your circuit is probably incomplete.

• Touching the probes one by one can help you check the continuity function of your multimeter. Now you can complete the circuit. Your multimeter should show you how to do this.

An electrical continuity test determines whether two points are electrically connected. If something is continuous, electric current can flow freely from one end to the other.

It is possible that there is no continuity. This may indicate a bad solder joint or a blown fuse or a circuit that is not wired correctly.

Replacing the Fuse

A common mistake with a new multimeter is measuring the current on the breadboard by probing from VCC to GND. This will immediately short the power to ground through the multimeter, causing the board to discharge. The internal fuse heats up when current flows through the multimeter. It then wears out and causes a 200mA loss. This happens within seconds and without any physical or audible indication that anything is wrong.

Note that the current measurement is taken in collection. To measure current, you need to break the VCC line to the breadboard or microcontroller. A blown fuse can cause the multimeter to read 0.00 and the system will not activate when the multimeter is connected. The internal fuse has been damaged and is used as a broken or broken cable.

Find your handy mini screwdriver and start attaching the screws to the fuse. The multimeter components and PCB tracks are designed to accept different current levels. If you accidentally push 5A to the 200mA port, your multimeter will be damaged or destroyed.

Sometimes you will need to measure high current tools such as an electric motor or a torch. Do you see the two areas where the red probe should be located on the front panel of your multimeter? 10A left and right. The fuse may blow if you try to measure more than 200mA using the mAV port. The risk of blowing a fuse is much lower if you use the 10A port for current measurement. Sensitivity is a trade-off. As mentioned above, the 10A port with the handle setting will only allow you to measure at 0.01A or 10.0mA. The 10A port and configuration works well for many systems that use more than 10mA. A 200mA port with 2mA, 200uA or 20uA may be the best choice if you need to measure very low power (micro and nano amplifiers).

Conclusion

You are now ready to use your digital multimeter to measure the world around you. You can immediately use it to answer many questions. Many questions about electronics can be answered with a digital multimeter.

Every electronics lab needs a multimeter as a basic piece of equipment. You may learn how to use a multimeter by reading this instruction. This guide will show you how to use a multimeter to measure voltage, current, and resistance. It also explains how to check for continuity.

www.ingramcontent.com/pod-product-compliance
Lightning Source LLC
Chambersburg PA
CBHW071459150726
48000CB00006B/2629